For Bobby, with zero reservations
— B.G.

To my number zero fan, Jocelyne, who was there
before anyone else
— J.C.

Text copyright © 2022 by Bruce Goldstone
Illustrations copyright © 2022 by Julien Chung

Library of Congress Cataloging-in-Publication Data Available
ISBN 978-1-338-74224-4
10 9 8 7 6 5 4 3 2 1   22 23 24 25 26
Printed in China 62
First edition, August 2022
Book design by Doan Buu
The text type was set in Avenir. The display type was set in Futura Condensed ExtraBold.
The illustrations were created digitally, using Adobe Illustrator, Photoshop, and Procreate.

ZERO
ZEBRAS

A Counting Book about
What's Not There

by **Bruce Goldstone**

illustrated by **Julien Chung**

Orchard Books
An Imprint of Scholastic Inc. | New York

I see one wallaby . . .

. . . and zero zebras.

Two tuna splish
and splash
and splosh . . .

. . . with zero zebras.

Three thrushes fly high in the sky . . .

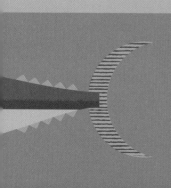

. . . by zero zebras.

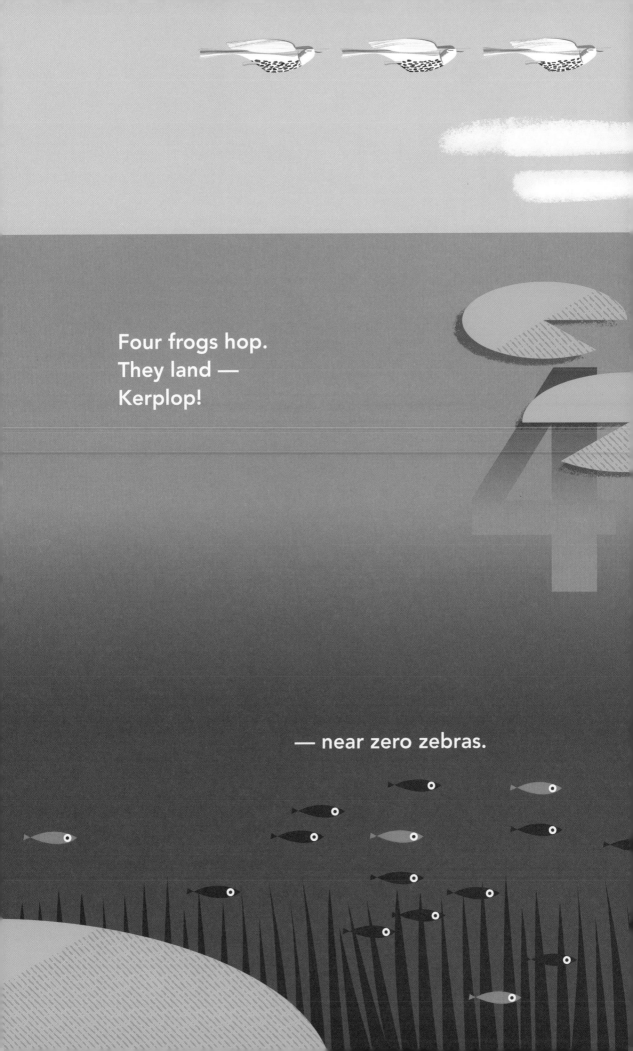

Four frogs hop.
They land —
Kerplop!

— near zero zebras.

Five foxes curl up
in their lairs . . .

. . . while zero zebras
curl in theirs.

Six seals are basking in the sun.

How many zebras bask
beside them?
Absolutely none.

Seven spiders spin
and in their webs
they pin . . .

. . . zero zebras.

Eight elephants trumpet
an elephant song . . .

. . . while zero zebras
sing along.

Nine newts creep
and crawl and slide . . .

. . . with zero zebras
by their side.

Ten tigers tiptoe —
that's how many.

What about zebras?
There aren't any.

Eleven llamas
like to spit.

It's zero zebras
that they hit.

Twelve turtles
wallow in the mud.
One is tucked
inside its shell.

Zero zebras wallow, too —
in zero zebra shells as well.

What's next? What's here?
What do you see
perching in this tree?
Why, look at that!
By now you've guessed.
Zero zebras — obviously.

But that's not all that isn't here!
Do you see zero eagles?
You'll find them next to zero pigs
and zero barking beagles.

What zeroes can you spy
in this summer sunset sky?

Zero beavers,
zero bats,
zero camels,
zero cats,
zero rhino,
zero rats,

When the stars come out tonight,
zero zebras do, too.
Along with zero pandas
eating zero bamboo.

Also zero penguins,
zero parrots, zero prawns.
Zero squirrels, zero sparrows,
zero snails, zero swans.

As well as zero cupcakes,
zero crayons, zero clocks,
zero scissors, zero slippers,
zero sailboats, zero socks.

So when you want to count a lot,
don't count what's there. Count what's not.

Try counting zeroes with your friends.
The list of zeroes never ends!

Some Thoughts About Zero

Zero is a big nothing, but it's also a big deal. Do you skip past zero when you start to count? Almost everybody does: 1, 2, 3, and more. But zero is always there first. Before there's something to count, there's zero, saying loud and proud that there's nothing there. But what isn't there?

Think about an empty box. What *isn't* inside? There are no shoes and no shirts. No books and no pencils. A yo-yo? A toaster? A roller coaster? Nope, nope, and nope. The list of what *isn't* inside the box goes on and on and on . . . and then on and on some more. In fact, there's literally no end to that list. You can always think of more things that aren't inside. There's a name for a list that goes on and on and on without ever reaching the end: infinity.

Infinity isn't really a number, it's an idea. What's the biggest number you can think of? A million? Infinity is bigger than that. A zillion? Infinity's bigger. A google is a 1 followed by 100 zeroes. That's a crazy big number, but infinity is crazy bigger. It's an idea because no matter how big a number you can count, infinity is always bigger.

The symbol for infinity looks like you gave a zero a twist in the middle (or maybe like two zeroes kissing).

Counting numbers are great for things you can see or feel or smell or taste or hear, but zero can name things you just think about. The possibilities go on forever. That's how zero can lead you into infinity. The only limit is your imagination.

BRUCE GOLDSTONE

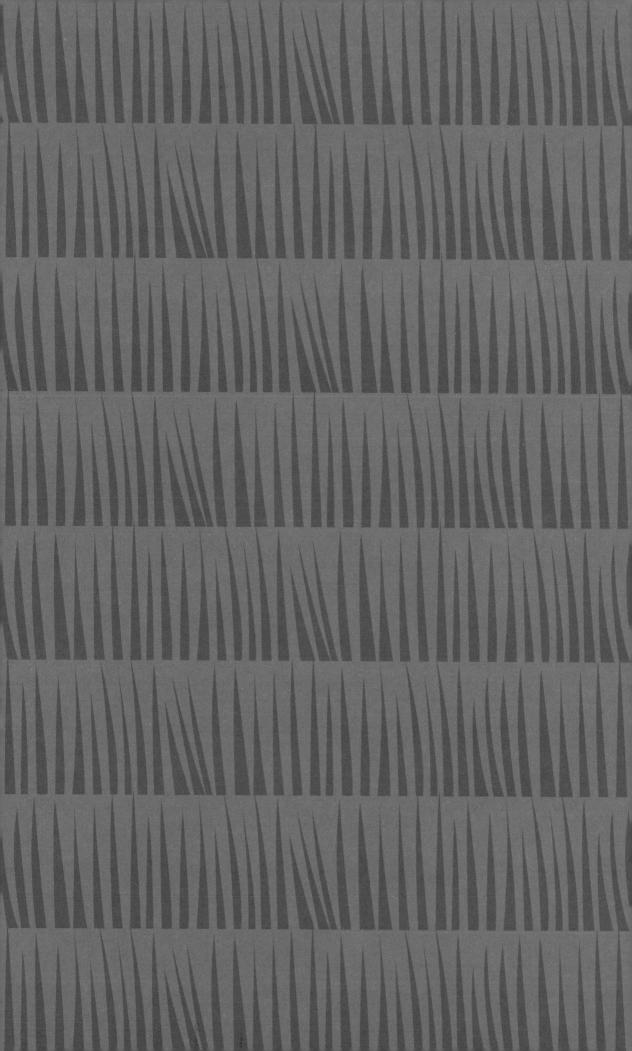